TAKS Review Every Day!
Practice Tests

HOLT, RINEHART AND WINSTON

A Harcourt Education Company

Austin · New York · Orlando · Atlanta · San Francisco · Boston · Dallas · Toronto · London

Table of Contents

Social Studies Practice Test 3

English Language Arts Practice Test 29

Mathematics Practice Test 41

Student Answer Sheet 53

Printed in the United States of America

ISBN 0-03-069077-3

1 2 3 4 5 6 7 8 9   82   07 06 05 04 03 02

SOCIAL STUDIES PRACTICE TEST

DIRECTIONS
Read each question and choose the best answer. Then mark the letter for the answer you have chosen.

SAMPLE A

Machine politics helped which U.S. president's grandfather become mayor of Boston?

A John F. Kennedy

B Jimmy Carter

C Richard Nixon

D Bill Clinton

STOP

Use the chart _and_ your knowledge of U.S. history to answer the following question.

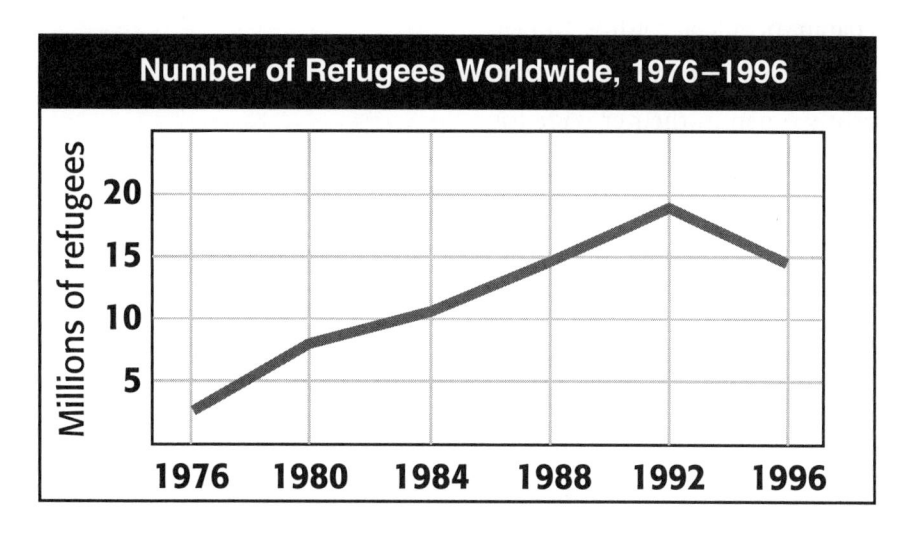

Number of Refugees Worldwide, 1976–1996

1   Which of the following tables correctly reflects the information in the chart?

A   Number of Refugees Worldwide

1976	5 million
1980	10 million
1984	10 million
1988	15 million
1992	20 million
1996	15 million

C   Number of Refugees Worldwide

1976	5 million
1980	4 million
1984	10 million
1988	15 million
1992	17 million
1996	15 million

B   Number of Refugees Worldwide

1976	15 million
1980	17 million
1984	15 million
1988	10 million
1992	7 million
1996	3 million

D   Number of Refugees Worldwide

1976	3 million
1980	8 million
1984	10 million
1988	15 million
1992	19 million
1996	15 million

GO ON

2 The "Jim Crow" laws were intended —

   F to separate people according to race.

   G as a way to prevent damage to crops.

   H to assist women in their crusade for the right to vote.

   J to remove barriers to employment for new immigrants.

GO ON

3   President Wilson helped prevent a national railroad strike by —

    A   allowing more frequent breaks to all workers.

    B   proposing that the workday be shortened to eight hours.

    C   allowing women to work on the railroad.

    D   adding children to the work force to solve the labor shortage.

4   Leading U.S. officials worried that the Treaty of Versailles —

    F   gave away too much to Germany and other defeated nations.

    G   was too complex to be implemented effectively.

    H   failed to include all nations involved in the war.

    J   was too harsh and humiliating to the defeated countries.

GO ON ▶

5 Which was NOT a reason for President Truman's decision to drop an atomic bomb on Japan?

A Japan's refusal to unconditionally surrender

B Japan's wartime atrocities

C It might lead to a global arms race.

D The saving of many lives on both sides

6 During World War II, Japanese Americans in the United States —

F were put into internment camps.

G assumed important roles in the war effort.

H protested in massive numbers.

J mostly migrated back to Japan.

7 The "final solution" was the German Nazi Party's goal of —

A eliminating poverty through public works programs.

B using V-2 rockets to attack the United States.

C exterminating the Jewish people.

D intercepting Allied communications.

8 In 1963 the March on Washington, D.C., led by Dr. Martin Luther King Jr. and others, helped focus attention on the —

F need to end the United States's involvement in Vietnam.

G discrimination problems facing African Americans.

H drought problems facing American farmers.

J unfair practices against unions and working people.

GO ON

9 During the 1980s and 1990s, scientists suggested that the thinning of the ozone layer was caused by —

   A natural processes that are unavoidable.

   B chemicals produced in the manufacture of sprays and coolants.

   C smoke from coal and wood fires.

   D damage caused by traditional methods of farming.

10 President Taft employed the "dollar diplomacy" policy in Latin America in order to —

   F hire skilled diplomats as U.S. representatives.

   G get South American nations to adopt U.S. currency.

   H use economic influence instead of military force.

   J increase U.S. military presence around the world.

GO ON ▶

11 Frustrations with the slow pace of the civil rights movement led to the emergence of black nationalism, which promoted —

  A a separate, self-reliant African American nation.

  B support for President Johnson's civil rights initiatives.

  C greater U.S. patriotism among African Americans.

  D nonviolent civil disobedience to change laws.

12 The main cause of the Red Scare in the United States in 1919 was —

  F increased dangers of a second world war.

  G marine pollution in U.S. fisheries.

  H the Bolshevik revolution in Russia.

  J the women's suffrage movement.

13 Lost Generation writers portrayed a certain kind of disillusionment in their reactions to —

  A the uselessness of war.

  B the lack of prosperity during the Jazz Age.

  C tax increases that burdened the poor.

  D losses of U.S. jobs to foreign workers.

GO ON ➡

Use the quote <u>and</u> your knowledge of U.S. history to answer the following question.

> You may well ask: "Why direct action? Why sit-ins, marches, and so forth? Isn't negotiation a better path? You are quite right in calling for negotiation. Indeed, this is the very purpose of direct action. Nonviolent direct action seeks to create such a crisis and foster such a tension that a community which has constantly refused to negotiate is forced to confront the issue. . . .
>
> For years now I have heard the word "Wait!" It rings in the ear of every Negro with piercing familiarity. This "Wait" has almost always meant "Never." We must come to see, with one of our distinguished jurists, that "justice too long delayed is justice denied."
>
>   — Martin Luther King Jr., "Letter from Birmingham Jail"

14 In 1962–63, Southern Christian Leadership Conference protesters adopted all of the following direct actions to improve conditions for African Americans EXCEPT —

   F boycotts.

   G marches.

   H sit-ins.

   J sabotage.

GO ON ▶

15 Which of the following is NOT true about Prohibition?

  A Prohibition laws helped encourage criminals to sell liquor illegally.

  B Prohibition laws were respected throughout the country.

  C Prohibition ended when the U.S. Constitution was changed.

  D Alcoholism and alcohol-related deaths declined during Prohibition.

16 Which two women were widely recognized as leaders in campaigns for women's rights?

  F Olivia Anderson and Lucy Griffith

  G Susan B. Anthony and Elizabeth Cady Stanton

  H Maggie Walker Green and Cecilia L. McDowell

  J Lydia R. Olson and Belinda Davis Wilson

17 The United States entered World War II after —

  A the Japanese bombed Pearl Harbor.

  B Italy and Germany invaded North Africa.

  C the British invaded Germany.

  D Germany attacked Poland.

GO ON ▶

18 Developing countries have been less able to benefit from advances in computer technology than the United States has because —

  F English is the only language that computers can recognize.

  G there has been little interest in computers outside the United States.

  H it is less expensive to mail or fax documents.

  J they lack adequate phone lines.

19 President Theodore Roosevelt promoted conservation by —

  A sponsoring the first recycling efforts.

  B protecting endangered species.

  C preventing the building of roads.

  D asking Congress to create national parks.

Use the information in the box <u>and</u> your knowledge of U.S. history to answer the following question.

> **Social Security Act**
> Provided unemployment insurance to workers and gave pensions to retired workers
>
> **Federal Deposit Insurance Corporation**
> Guaranteed bank deposits
>
> **Civilian Conservation Corps**
> Trained workers to plant trees and develop recreation areas
>
> **Public Works Administration**
> Created jobs by authorizing the building of roads, buildings, and other projects to stimulate business activity

20 Which of these New Deal programs continues to play a central role in the national political debate?

  F Federal Deposit Insurance Corporation

  G Social Security Act

  H Civilian Conservation Corps

  J Public Works Administration

GO ON ▶

21 *Brown* v. *Board of Education*, the landmark 1954 Supreme Court decision, was important because it helped establish —

A the interstate highway system.

B women's athletic programs in universities.

C the separation of church and state.

D the illegality of segregated schools.

Use the poem to the right _and_ your knowledge of U.S. history to answer the following question.

22 The poem illustrates which of the following difficulties U.S. forces had in achieving a military victory in Vietnam?

F U.S. soldiers found it difficult to identify and defeat a guerrilla army.

G The North Vietnamese had more powerful weapons.

H The United Nations occupied Vietnam as a peacekeeping force.

J The U.S. military was underequipped and undermanned.

"Guerrilla War"
by William D. Ehrhart

It's practically impossible
to tell the civilians
from the Vietcong.

Nobody wears uniforms.
They all talk
the same language
(and you couldn't under-
    stand them
even if they didn't).

They tape grenades
inside their clothes,
and carry satchel charges
in their market baskets.

Even their women fight;
and young boys,
and girls.

It's practically impossible
to tell civilians
from the Vietcong;
after awhile
you quit trying.

GO ON

Use the time line <u>and</u> your knowledge of U.S. history to answer the following question.

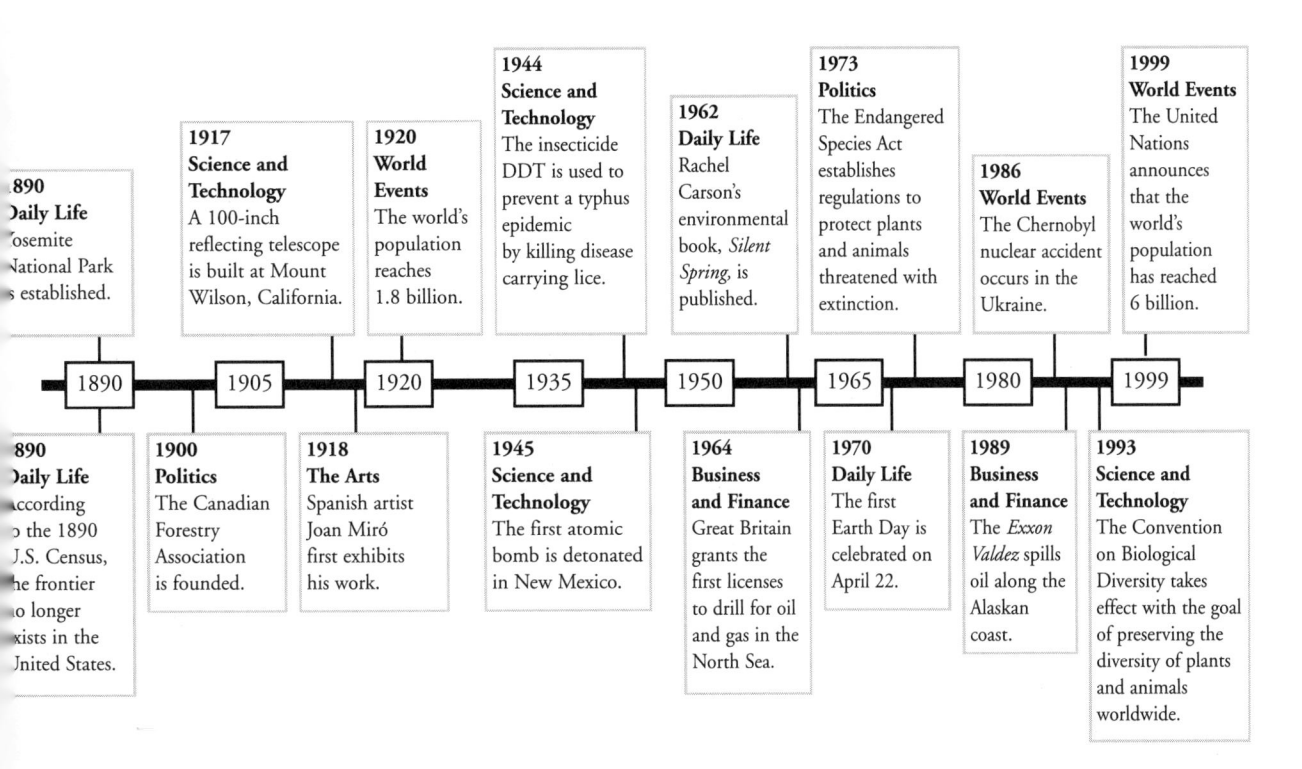

1890
Daily Life
Yosemite National Park is established.

1917
Science and Technology
A 100-inch reflecting telescope is built at Mount Wilson, California.

1920
World Events
The world's population reaches 1.8 billion.

1944
Science and Technology
The insecticide DDT is used to prevent a typhus epidemic by killing disease carrying lice.

1962
Daily Life
Rachel Carson's environmental book, *Silent Spring*, is published.

1973
Politics
The Endangered Species Act establishes regulations to protect plants and animals threatened with extinction.

1986
World Events
The Chernobyl nuclear accident occurs in the Ukraine.

1999
World Events
The United Nations announces that the world's population has reached 6 billion.

| 1890 | 1905 | 1920 | 1935 | 1950 | 1965 | 1980 | 1999 |

1890
Daily Life
According to the 1890 U.S. Census, the frontier no longer exists in the United States.

1900
Politics
The Canadian Forestry Association is founded.

1918
The Arts
Spanish artist Joan Miró first exhibits his work.

1945
Science and Technology
The first atomic bomb is detonated in New Mexico.

1964
Business and Finance
Great Britain grants the first licenses to drill for oil and gas in the North Sea.

1970
Daily Life
The first Earth Day is celebrated on April 22.

1989
Business and Finance
The *Exxon Valdez* spills oil along the Alaskan coast.

1993
Science and Technology
The Convention on Biological Diversity takes effect with the goal of preserving the diversity of plants and animals worldwide.

23 What would be an appropriate title for this time line?

A Changes in World Population During the Twentieth Century

B Environmental Issues of the Twentieth Century

C Environmental Disasters of the Twentieth Century

D The U.S. Government's Role in Environmental Issues During the Twentieth Century

GO ON

24 Franklin D. Roosevelt won the election of 1932 because —

F he was a better debater than his opponent.

G he was an inspiring wartime leader.

H he was a popular incumbent president.

J voters hoped he would solve the country's economic problems.

25 Why did more women join the work force during the early 1940s?

A To help end the Great Depression

B To replace men who went to war

C Because of the Equal Rights Amendment

D As a result of advertising for job openings

26 The United States built the Panama Canal in order to —

F cut the travel time between the Atlantic and Pacific Oceans.

G connect North and South America with railroads.

H develop South American industry.

J restore democracy to Colombia.

GO ON

Use the graph <u>and</u> your knowledge of U.S. history to answer the following question.

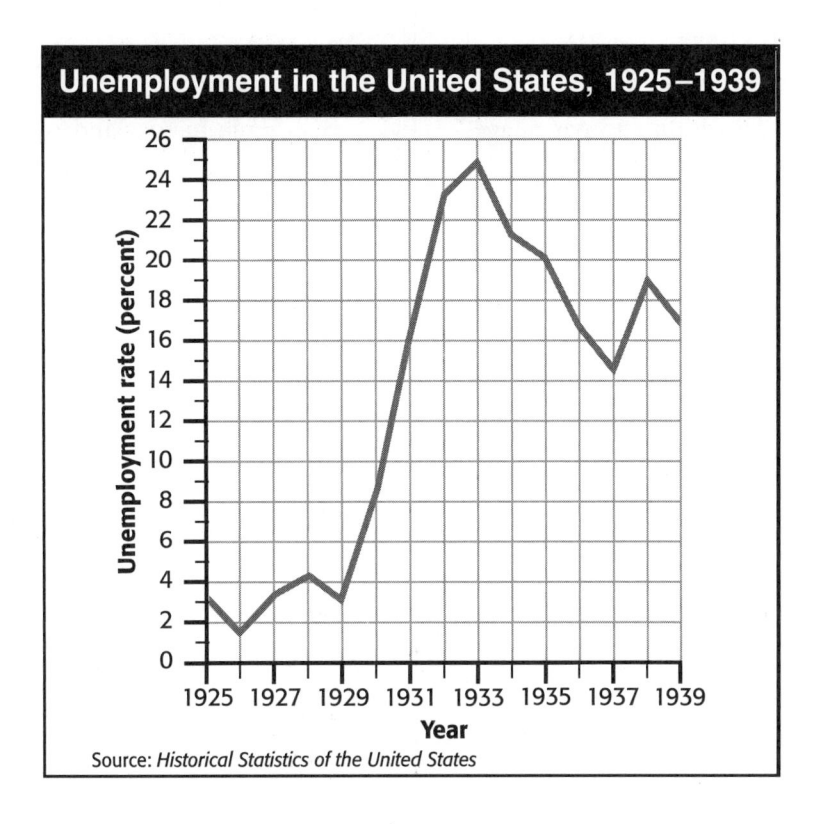

Unemployment in the United States, 1925–1939

Source: *Historical Statistics of the United States*

27 Which of the following statements can be supported by information in the graph?

A The economic policies of the New Deal had no effect on unemployment.

B New Deal policies caused a large jump in unemployment.

C The lingering effects of the Great Depression had passed by 1938.

D Unemployment remained a serious problem but the New Deal helped to reduce it.

GO ON ➤

28 The creation of the Civil Service Commission was part of a national effort in the late 1800s to —

F promote civil rights.

G eliminate corruption and patronage.

H protect the public from natural disasters.

J establish a military draft.

29 Writers and artists of the Harlem Renaissance created works about —

A African American identity and culture.

B Dutch history and culture.

C the experiences of European immigrants.

D the experiences of the Lost Generation.

GO ON ▶

Use the chart __and__ your knowledge of U.S. history to answer the following question.

New Deal Programs

Banks	Farmers	Labor	Business
• Emergency Banking Act • Federal Deposit Insurance Corporation	• Farm Credit Administration • Agricultural Adjustment Administration	• National Labor Relations Act • Fair Labor Standards Act	

30 Which programs or congressional acts accurately complete this chart?

   F Federal Emergency Relief Administration

   G Soil Conservation Service

   H Securities and Exchange Commission

   J Revenue Act of 1935

GO ON ▶

Use the passage <u>and</u> your knowledge of U.S. history to answer the following question.

> When I was sixteen, I was supposed to marry a man in Italy, but I didn't want him. My mama tell me, "Either you marry this guy or you go to America." But I told her, "I don't like him." she say, "Then you go to America." That's why I came to America. There was nothing in Italy, nothing in Italy. That's why we came. To find work, because Italy didn't have no work. Mama used to say, "America is rich, America is rich."
>
> — Carla Martinelli

31 What did Carla's mother mean by using the phrase "America is rich…"?

   A America is filled with gold.

   B America is a land rich with opportunity.

   C America is wealthy.

   D America is full of rich people.

32 The Highway Act of 1956 created an interstate highway system that led to —

   F the growth of suburbs.

   G the growth of cities.

   H an expansion of mass transit.

   J the isolation of small towns.

GO ON ▶

Use the chart <u>and</u> your knowledge of U.S. history to answer the following question.

U.S. Battle Deaths in Vietnam, 1961–1970
(approximate numbers)

Year	Total
1961–1965	1,500
1966	4,700
1967	9,600
1968	14,300
1969	6,700
1970	7,100

33 Which of the following bar graphs correctly reflects the information in the chart?

A

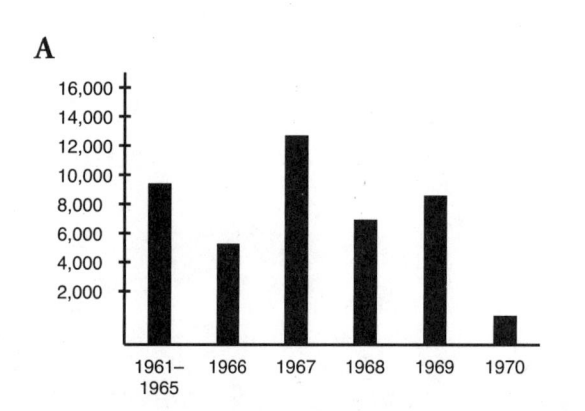

C

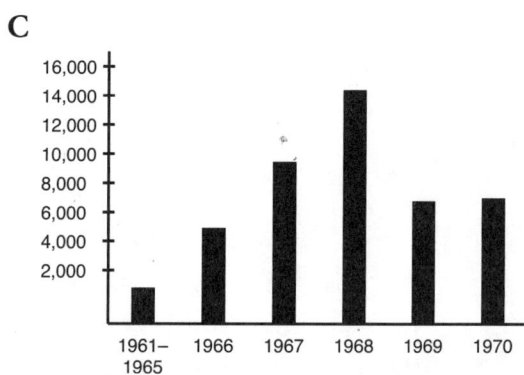

B

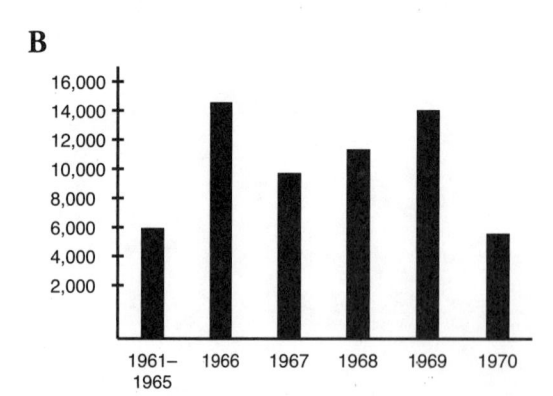

D

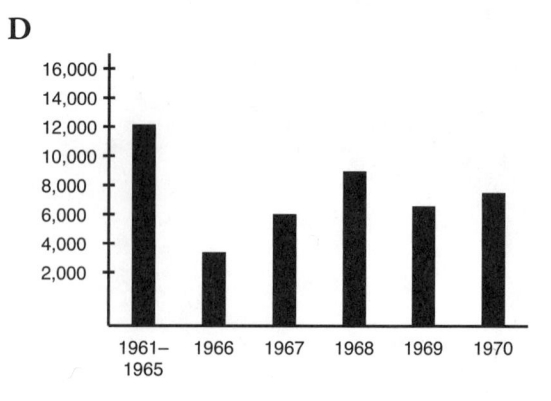

Use the information in the box <u>and</u> your knowledge of U.S. history to answer the following question.

The U.S. Economy in the 1950s

- More than 60 percent of Americans earned a middle-class income by the mid-1950s.

- Union membership peaked at 18.5 million in 1956.

- During the Eisenhower administration, Social Security and unemployment benefits were increased.

- The minimum wage was increased by one third in 1955.

34 Based on the information in the box, what conclusion can be drawn about the 1950s?

  F It was a time of political and economic turmoil.

  G Union activity adversely affected the economy.

  H It was a time of abundance for most people.

  J President Eisenhower was extremely unpopular.

GO ON

Use the chart <u>and</u> your knowledge of U.S. history to answer the following question.

Changes in Higher Education, 1950–1997

	1950	1997
Number of Institutions of Higher Education	1,863	4,064
Annual Student Body Enrollment	2,281,000	14,350,000
Annual Number of Bachelor Degrees Conferred	496,874	1,172,000
Percentage of Female Students	32%	57%

35 Which of the following best explains the changes shown in the chart?

A Farms grew larger and employed more unskilled laborers.

B Employers now more often require college training than they used to.

C More people are self-employed in the modern economy.

D There are now fewer working women.

GO ON

Use the cartoon <u>and</u> your knowledge of U.S. history to answer the following question.

© R.R. Lurie, Cartoonews International.

36 The cartoon suggests that President Richard Nixon faced a challenge in —

  F avoiding relations with China and the USSR.

  G balancing the competing interests of China and the USSR.

  H balancing the environmental issues between China and the USSR.

  J avoiding a summit meeting with China and the USSR.

GO ON ▶

37 The term the *Bessemer process* means —

A building bridges with steel.

B burning off steel impurities.

C fur trapping with steel traps.

D molding steel with heat.

38 The United States initiated which of the following in response to the devastation of European cities after World War II?

F United Nations

G NATO Alliance

H Truman Doctrine

J Marshall Plan

GO ON

Use the graph <u>and</u> your knowledge of U.S. history to answer the following question.

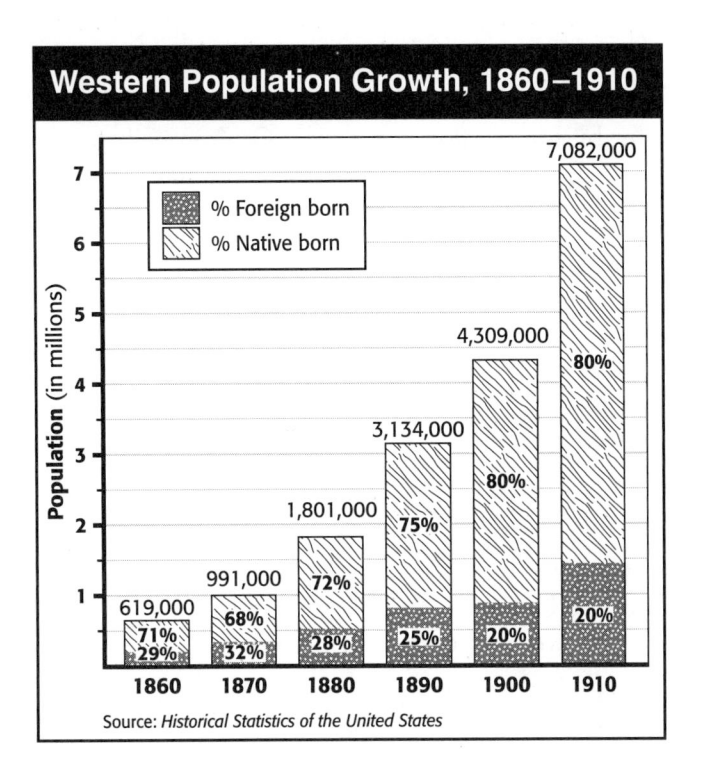

Western Population Growth, 1860–1910

Population (in millions)

% Foreign born
% Native born

619,000 — 71% / 29%
991,000 — 68% / 32%
1,801,000 — 72% / 28%
3,134,000 — 75% / 25%
4,309,000 — 80% / 20%
7,082,000 — 80% / 20%

1860   1870   1880   1890   1900   1910

Source: *Historical Statistics of the United States*

39 According to the graph, from 1860 to 1910, the numbers of foreign-born people in the West —

  A declined steadily.

  B stayed the same.

  C declined sharply.

  D increased sharply.

GO ON ▶

40 Cattle drives affected the settlement of the western United States in the mid-1800s by —

  F creating cattle towns where drives ended at railroad stops.

  G creating a need for businesses along the route.

  H establishing the need for colleges and universities.

  J decreasing the need for manufacturing.

41 Which event during the Civil War ensured the survival of the Union and weakened the power of the southern states?

  A General Sherman captured and burned Savannah.

  B President Lincoln declared that the slaves were free.

  C Mary Boykin Chesnut wrote to President Lincoln.

  D Generals Lee and Johnston surrendered.

42 By the 1880s and 1890s, what was one thing the U.S. government attempted to do with American Indians?

  F It tried to wipe out the American Indian population.

  G It hoped to assimilate American Indians into mainstream life.

  H It worked to convert American Indians to Christianity.

  J It preserved the American Indian way of life on reservations.

ENGLISH LANGUAGE ARTS PRACTICE TEST

Sophie has written this report for a history project. She has asked you to read her report and think about any suggestions you would make. When you finish reading the report, answer the following multiple-choice questions.

Why Come to America?

1   People come to the United States for many reasons. Since the 1960s, many of the

2   Asian and Latin American immigrants to the United States have been refugees. As

3   communist North Vietnam took over South Vietnam in 1975, hundreds of thousands of

4   them fled their country in crowded boats. Other refugees fled fighting in Cambodia and

5   Laos. The U.S. government, which had supported South Vietnam's war, created programs

6   to allow those refugees to escape from Communism and settle in the United States.

7   Thousands of refugees have also fled violent or opposite governments in Latin

8   American nations to come to the safety of the United States. The United States's reaction

9   to these refugees has been mixed. For many years the official U.S. policy was to accept

10 Cuban refugees from the communist state. This changed when he threatened to flood

11 southern Florida with thousands of refugees who wanted to escape Cuba's poverty.

12 But times have changed. Most recent immigrants have come seeking economic

13 opportunities. For example, many Mexican immigrants work for wages to support family

14 members back home. Why come to America? For freedom, for prosperity, for

15 many things.

16 Florida's former governor, Lawton Chiles, declared, "Florida cannot stand another

17 influx." The U.S. Coast Guard even began stopping the Cuban refugees at sea and

18 sending them to camps outside the United States. Eventually, however, the U.S.

19 government decided to resettle most of the refugees in the United States.

1 What is the **BEST** change, if any, to make in the sentence in line 1? (*People come…many reasons.*)

  A Insert a period after States

  B Change *come* to **immigrate**

  C Delete the period after *reasons*

  D Make no change

2 The meaning of the sentence in lines 2–4 *(As communist North Vietnam…in crowded boats)* can be improved by changing **them** to —

  F Americans.

  G Chinese.

  H Vietnamese.

  J Cubans.

3 Which phrase would **BEST** describe the United States's position based on lines 5–6?

  A Against the Communists

  B With North Vietnam

  C Against refugees

  D With the Russians

4 What is the **BEST** change, if any, to make in the sentence in lines 7–8? (*Thousands of refugees…of the United States.*)

  F Delete the word *safety*

  G Insert a comma after *thousands*

  H Change *opposites* to **oppressive**

  J Make no change

5 Which of these sentences would **BEST** fit with the ideas in Paragraph 2 (lines 8–9)?

  A Sometimes the United States would welcome immigrants and sometimes it would deport them.

  B The United States often deported immigrants arriving from Cambodia.

  C Latin American nations often protested the United States's strict policies on accepting immigrants.

  D Many people fled their homes when they knew of the immigrants coming.

GO ON ▶

6 Which transition should be added to the beginning of the sentence in lines 9–10? *(For many years…communist state.)*

    F For example,

    G Sometimes,

    H Not often,

    J Furthermore,

7 What is the **BEST** change, if any, to make in the sentence in lines 10–11? *(This changed when he…to escape Cuba's poverty.)*

    A Insert an exclamation mark after *poverty*

    B Delete *he*

    C Change *he* to **Fidel Castro**

    D Make no change

8 What is the **BEST** way to improve the conclusion?

    F Move Paragraph 2 to the end of the report.

    G Delete Paragraph 4

    H Revise Paragraph 1

    J Exchange Paragraph 3 with Paragraph 4

9 Which of these phrases would **BEST** fit with the ideas in lines 12–13? *(Most recent immigrants…economic opportunities.)*

    A To the Moon

    B To the United States

    C To Cuba

    D To Latin America

10 What is the **BEST** addition to the end of Lawton Chiles statement to clarify whom he is talking about in Paragraph 4 (lines 16–17)?

    F Of Americans

    G Of civilians

    H Of refugees

    J Of Cubans

11 What is the **BEST** change, if any, to make in the sentence in lines 18–19? *(Eventually, however…in the United States.)*

    A Delete *in the United States*

    B Add a comma after *the*

    C Change **refugees** to *refuges*

    D Make no change

GO ON ➡

DIRECTIONS

The following section is an assessment of your reading ability. Read the selection. Then read each question and all its options. Decide which option BEST completes or answers the question. Mark the letter for that answer on your answer document or write your answer in the space provided.

1968: Frederick Williamson

Interview and story by Sheila Nippo

This story is based on one of a series of interviews conducted by South Kingstown (RI) High School students in the Spring of 1998. All of the interviews were focused on recollections of the year 1968.

1   "We have won and are winning many battles, but we have not yet won the war. We have come a long way since the Sixties and much of our thinking today can be attributed to that period in time."

2   Things went along quite well when I was growing up. Lowell, Massachusetts was a thriving mill town during the Twenties. My family consisted of myself, my mother, my father, my two brothers, and my sister. When I was growing up, everything had to do with cowboys. All youngsters were interested in cowboys. I spent most of my time at the library. I read everything I could get my hands on. As a result of my reading, I always wanted to write. I wanted to be an author and write stories and books.

3   I remember my father as being a dyed-in-the-wool Republican. The rest of his family were Republicans as well. At that time, most African Americans were Republican. When Franklin D. Roosevelt came along during the Depression, many African Americans focused on his activities and joined in with the Democratic Party. I lived in Lowell until my mother and father separated in 1929, when I was fourteen years old. I moved to East Providence and lived with my uncle and then later I moved to Providence. The rest of my family stayed in Lowell for two more years before moving to East Providence.

My notes about what I am reading

GO ON ▶

My notes about what
I am reading

4   Throughout my life, I have had many experiences
with racism and discrimination because of the color of
my skin. Many times, I have been out searching for
apartments and been turned down. . . . I remember
sitting in restaurants in Providence waiting for service.
I would sit at the corner of the counter and wait to be
served. After a long wait, someone would finally come
over to me and say, "We don't serve colored people
here sir." I could have sat there forever and would have
never been served.

5   When I worked at the Naval Air Station, I was a
manager in a pining branch, in a supply department
and I had to learn all sorts of plans and procedures.
When I did get to write, it was mostly plans and
procedures, information for aircraft squadrons on
the base and other activities. Due to my parent's
separation, I had no choice but to go straight out
into the work force. In 1963, I helped to form the
Providence Human Relations Committee. I was
Vice Chairman and then was later Chairman of
the committee for ten years. I was very active in
attempting to see to it that all citizens received
equal treatment.

6   In 1963, when the very famous march in
Washington took place, most of the public, including
myself and others, were waiting to see what would
happen. A lot of my involvement with the Civil Rights
issues has been with fair housing. Much of the
reasoning was that people felt that blacks living in their
neighborhood was a bad thing. They all thought that
having a person of color living on their street would
bring down the value of their property. They thought
that if black folks lived in their neighborhood they
would be associating with their sons and daughters and
that would motivate them to marry each other.
That would be the worst possible thing that anyone
could think of.

GO ON ➤

7   I was very happy when the schools were desegregated. To take young people and separate them into separate little islands is morally and ethically wrong. It provided a disservice to the nation. I feel that you can get to know those who are good and those who aren't so good and understand that there are good and bad people in every race that exists in the world.

8   During Malcolm X's rise to power, he and I did not agree on many things. I am more conservative than he. Despite my not agreeing with him, I have to appreciate the fact that somewhere along the line there has to be someone to stand up for balance and power. I appreciated him for what he did and what he represented. Then, there were the Black Panthers. I was not in favor of the Black Panthers. They adopted an aggressive attitude and role that woke the people up to the dangers of this kind of segregation. I said to the public, "This is the kind of confrontation that will eventually happen unless you, America, wake up to the fact that everyone is to be treated equally." The more passive and laid back America was, the longer discrimination and segregation would exist, fueling the Black Panthers' fire.

9   Martin Luther King Jr.'s death was one of sorrow and distress. The assassination of King was one of the most terrible incidents that can be looked back on in American history. He was the kind of person who only comes around once in a lifetime or once in several lifetimes. His thoughts and the way he served, his ability to lead in a nonviolent way was one that was very much appreciated. It was a loss not just for America but a loss for the world. It wasn't just a loss for minorities such as African-Americans, it was a loss for whites too. We needed a voice of reason who was able to appeal to people's sense of what was right, to say things publicly. It was a very sad thing. When the news came of his assassination, I received a call from the pastor of the Grace Episcopal Church in Providence. They were holding a service the day following the assassination. The next day, I participated in the service. I gave a homily stating what he had meant to us and how, at that particular time, his assassination was an important point in my life.

GO ON ▶

10   John F. Kennedy was an interesting, personable man who had an interesting war record, serving in a PT boat squadron. He excited the imaginations of most of the people in the United States. When he was elected president, he spoke those words, "Ask not what your country can do for you, but what you can do for your country." It was such a remarkable phrase. People were just waiting for something like this. It was one of the great phrases that touched our imaginations. It was a very sad day when he was assassinated. John Kennedy brought to America and to the government a totally new look that made people feel good about being an American.

11   What the Sixties accomplished was getting people's attention. It was a time when people were allowed to get up and voice their thoughts. People were allowed to have the voice and the decision making that now have allowed them to have a better feeling. They had done something to make a difference. These days, you can go down South and find many cities and towns with black mayors. There is a lot of political power that has been taken over by minorities. That is how it should be. There are a lot of boarding rights now accepted throughout the south.

12   As far as minorities are concerned, the Sixties helped to place tension on the fact that it was a man's world. Women were treated as second-class citizens like the minorities were. We now have a better sense and feeling of community, but many of the goals of the Sixties still have not been achieved. Will the goals ever be reached? Human nature is a very dynamic thing. In the Sixties there were regulations passed. Fair housing and women achieving balance and recognition in matters of employment and other areas of interest. A number of things came out of the Sixties that gave Americans a good hard look at themselves. It is something that we still have to work at. Today it is understood, getting a few pieces of legislation passed is not the end. We have won and are winning many battles, but we have not yet won the war. We have come a long way since the Sixties and much of our thinking today can be attributed to that period in time.

12 According to the selection, the author's childhood in Lowell, Massachusetts, was —

  F troubled and unhappy.

  G an ongoing struggle for civil rights.

  H spent reading and thinking about cowboys.

  J something he would rather keep private.

13 As a youth, what did he hope to do when he grew up?

  A Become a civil rights leader

  B Become an astronaut

  C Become a firefighter

  D Become an author

14 The main idea of Paragraph 4 is that the author recalls —

  F experiences of racial discrimination.

  G meals eaten at lunch counters.

  H trying to rent an apartment.

  J waiting a long time for service.

15 What evidence do you see that the author was interested in using his writing abilities in his working life?

  A He was quick to join the work force when his parents separated.

  B There is no evidence given in this selection.

  C He had produced many unpublished novels and plays.

  D He wrote plans and procedures at the Naval Air Station.

16 The main idea of Paragraph 8 is that the author —

  F agreed with the ideas of Malcolm X and the Black Panthers.

  G disagreed with Malcolm X but agreed with the Black Panthers.

  H disagreed with both Malcolm X and the Black Panthers, but felt they had an impact on the issue.

  J was impatient with the conservatism of Malcolm X.

17 According to Paragraph 9, Dr. Martin Luther King Jr. was appreciated —

  A for his skills as a doctor.

  B as someone who stirred up anger.

  C as a voice of reason.

  D to this day.

18 The purpose of Paragraph 11 is —

  F to show that some African Americans had become mayors.

  G to point out the lasting importance of the Sixties.

  H to introduce some unfamiliar vocabulary.

  J to show that little has changed since the Sixties.

19 The author's attitude about the Sixties can BEST be described as —

  A resentful.

  B hysterical.

  C humorous.

  D appreciative.

GO ON ▶

20 The author uses a metaphor in Paragraph 1 to describe —

   F the Civil War.

   G the war in Vietnam.

   H the Battle of Midway.

   J the struggle for equality and fairness for African Americans.

21 According to the author, when he was a child, most African Americans —

   A were Democrats.

   B were Republicans.

   C were Black Panthers.

   D were not affiliated with any political party.

22 According to Paragraph 6, the issue that inspired the author to become involved in civil rights was —

   F school integration.

   G "whites-only" public accommodations.

   H fair housing policies.

   J hiring discrimination.

23 The author admires John F. Kennedy for the way he —

   A inspired people to be proud of their American ideals.

   B led the fight for civil rights.

   C fought in World War II.

   D was elected president.

24 The purpose of Paragraph 12 is to —

   F amuse and entertain.

   G set the tone for the descriptions that follow.

   H reflect on a series of disasters.

   J sum up and evaluate a time gone by.

25 From the author's account, a reader might predict that he would eventually —

   A become an officer in the Air Force.

   B forget about the controversies of the Sixties.

   C become a writer.

   D become a real estate broker.

26 Compared with Dr. Martin Luther King Jr., Malcolm X was more —

   F happy-go-lucky about his experiences.

   G determined for African Americans to exercise power.

   H committed to civil disobedience and nonviolence.

   J unconcerned with what was happening in the Northeast.

27 Do you think the author believes the controversies of the Sixties lead to positive or negative changes? Explain your answer.

_____

_____

_____

_____

_____

_____

_____

_____

_____

_____

_____

 GO ON

WRITING TASK

The following activity is designed to assess your writing ability. The prompt will ask you to explain something. You may think of your audience as being any reader other than yourself.

When scorers evaluate your writing, they will look for evidence that you can:

- respond directly to the prompt;

- make your writing thoughtful and interesting;

- organize your ideas so that they are clear and easy to follow;

- develop your ideas thoroughly by using appropriate details and precise language;

- stay focused on your purpose for writing by making sure that each sentence you write contributes to your composition as a whole; and

- communicate effectively by using correct spelling, capitalization, punctuation, grammar, usage, and sentence structure.

WRITING PROMPT

In *The Whole World Was Watching: An Oral History of 1968*, a participant in the social changes of the era reflects on the ways that conflict led to improved conditions for African Americans.

Write a persuasive essay in which you discuss the idea of whether the civil rights movement was successful in achieving its goals. Provide reasons to support your argument with explanations or illustrations from your knowledge of U.S. history.

GO ON ▶

SAMPLE RESPONSE

Was the civil rights movement of the 1960s successful in achieving its goals? The answer depends on who you ask. If we judge its success by the goals set out by Dr. Martin Luther King Jr. in his famous speech given during the march on Washington, we know that a great many things were accomplished, though a significant number remain incomplete.

King said that he hoped for a day when African Americans are "judged not by the color of their skin but by the content of their character." He also hoped for a day when freedom and justice would be available to all people of all faiths and races throughout the entire United States.

These are the highest goals that a society can aspire to, and they represent an ideal. In his speech, Dr. King sought the eradication of discrimination, racism, and hate, which held African Americans back from opportunities and limited many to poverty. Because of the efforts of many brave people of all races, many of the legal barriers to equal opportunity were dismantled during the years following this speech.

Some of the significant changes that resulted from the civil rights movement include the Voting Rights Act, which helped ensure a greater voice to African Americans, along with legislation that also helped limit and deter discrimination in housing, employment, and access to education. These were some of the most important victories of the civil rights movement. In addition to changes in federal and state laws, courts also helped strike down laws and practices that were too harsh and discriminatory.

Despite these victories, the goals of freedom and justice that Dr. King called for in his speech did not come for all Americans. Antidiscrimination laws could remove some of the barriers that held people down, but they would not automatically raise poor people out of poverty. They could not completely change the habits and attitudes of people, both privileged and underprivileged, that had developed since America was first colonized at Jamestown in 1607.

In the twenty-first century, racial discrimination continues to exist, though the government no longer gives official legal approval to it. These advances have allowed millions of people to participate more fully and prosperously in American society. However, the advances have not eliminated the problems that Dr. King spoke of, and additional work remains to realize his dream.

MATHEMATICS PRACTICE TEST

DIRECTIONS

Read each question and choose the best answer. Then mark the letter for the answer you have chosen. If a correct answer is <u>not here</u>, mark the letter for "Not Here."

1 Which of the following relations is NOT a function?

   A {(5, -2), (-2, 5), (-4, 0), (0, -4)}

   B {(3, -3), (-1, 3), (-3, 6), (5, -5)}

   C {(2, 7), (-4, 6), (-1, 2), (2, -4)}

   D {(8, 3), (-2, -2), (4, 5), (2, -2)}

2 The table gives the input and output values for a function.

x	1	2	3	4
f(x)	9	8	7	6

What is the function rule for the table?

   F $x + 8$ **H** $x + 1$

   G $10x - 1$ **J** $10 - x$

3 It costs $3 for the first hour and $1 for each additional hour to park a car in a parking garage. If you park for *n* hours, which function describes this relationship?

   A $3n$ **C** $3n + 1$

   B $n + 3$ **D** $n + 2$

4 The base of an isosceles triangle is 12 centimeters long and each of the legs is 10 centimeters long.

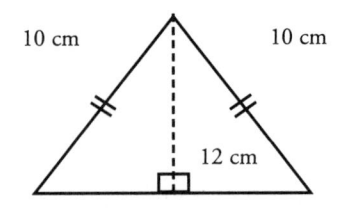

What is the height of the triangle?

   F 6 cm   **H** 8 cm

   G 7 cm   **J** 9 cm

5 Melinda wants to cover the front of her sticker album with stickers so that none of the stickers overlap and there are no gaps. Which shape should she use to do this?

   A   **C**

   B   **D**

GO ON

6 A surveyor needs to find the width of the lake.

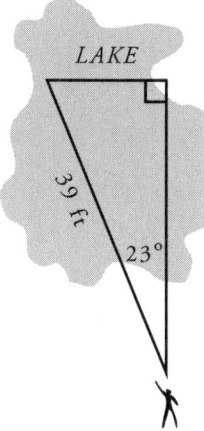

Which equation can be used to find the width of the lake?

F tan 23 = x/39   **J** $x^2 + 23^2 = 39^2$

G cos 23 = x/39   **K** Not Here

H sin 23 = x/39

7 Which scatterplot would probably show a positive relationship?

A age, height

B age, hair color

C age, amount of TV watched

D hair color, height

E hair color, amount of TV watched

8 A cereal company is designing a new box. If the height of the box is 11 inches, the width is 2 inches and the length is 8 inches, how much cardboard is needed for the box?

F 126 in.^{2} **H** 236 in.^{2}

G 176 in.^{2} **J** 252 in.^{2}

9 Which expression is equivalent to $3(2n + 7) - 5n - 9$?

A $n + 12$ **C** $n - 12$

B $11n - 30$ **D** $11n + 30$

10 A 15-foot ladder is placed against the house. As the bottom of the ladder is moved farther away from the house, what happens to the slope of the ladder?

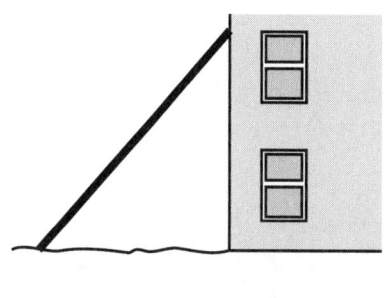

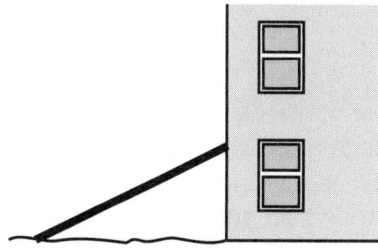

F The slope remains the same.

G The slope increases.

H The slope decreases.

J The slope cannot be determined.

11 What is the y-intercept for $2x + y = 6$?

A 0   **C** 3

B 32   **D** 6

GO ON

43

12 The diagram shows △PQR and line PS.

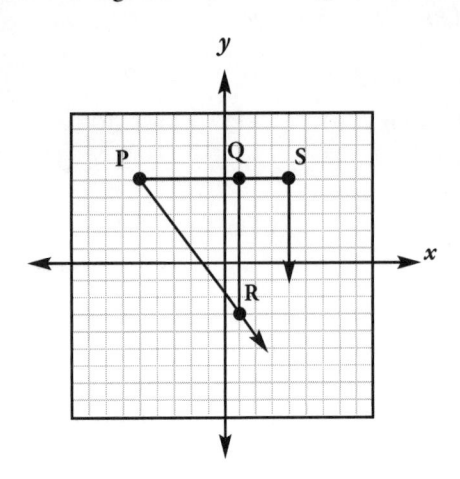

Where should point T be located on the grid so that △PQR and △PST are similar?

F (4, -6)   **H** (3, -5)

G (4, -7)   **J** (2, -4)

13 In a survey of favorite ice cream flavors, 30 of the 50 people surveyed preferred chocolate. In a group of 10,000 people, how many would you expect to prefer chocolate ice cream?

A 30   **C** 3,000

B 600   **D** 6,000

14 Five runners competed in a race. Nate beat Pam. Jake was not last. George was beaten by Tina and Jake. Jake crossed the finish line just after Tina did. Tina lost to Pam. In what order did the friends finish the race?

F Nate, Pam, Tina, Jake, George

G Jake, Nate, Pam, Tina, George

H Nate, Pam, Jake, Tina, George

J Tina, Jake, George, Nate, Pam

15 What are the next two numbers in this pattern?

1, 1, 2, 3, 5, 8, 13, …

A 20, 29   **C** 21, 34

B 17, 19   **D** 18, 23

16 The solid figure is built with cubes. Which could represent the shape of the solid figure when viewed from directly above?

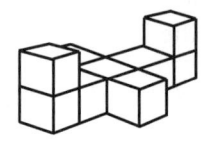

F   **H**

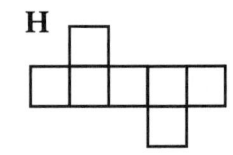

G   **J**

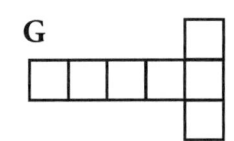 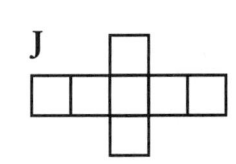

17 Which of the following patterns can be used to form a cube?

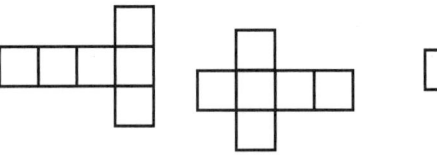

      I   II   III

A I and II only   **C** I and III only

B II and III only   **D** I, II, and III

GO ON ➤

18 A flagpole is 15 feet tall. A guy wire is attached to the top of the flagpole to a point 8 feet from the base of the flagpole.

What is the length of the guy wire?

F 23 ft   **H** 20 ft

G 21 ft   **J** 17 ft

19 At Foodmart, 3 pounds of tomatoes and a head of lettuce cost $5.36. The cost of 2 pounds of tomatoes and 2 heads of lettuce is $5.16. What is the cost of a head of lettuce?

A $0.99   **C** $1.29

B $1.19   **D** $1.39

20 Mr. Rodriguez wants to make a rectangular garden that has an area of 36 square feet. The length of the garden must be 5 feet more than the width. Which quadratic equation can be used to find the width of the garden?

F $x^2 - 36 = 0$

G $x^2 - 5x + 36 = 0$

H $x^2 + 5x + 36 = 0$

J $x^2 + 5x - 36 = 0$

21 Lila wants to leave a tip of about 15 percent on her restaurant bill of $60.75. About how much should she leave for a tip?

A $8   **C** $10

B $9   **D** $11

22 Pat can choose from 4 tops, 5 bottoms, and 2 pair of shoes for her school uniform. What strategy could you use to find out how many possible outfits she has?

F Make a list

G Guess and check

H Work backwards

J Work a simpler problem

23 A cable television company charges $15.95 a month for basic cable service. Each premium channel costs an additional $3.95 per month. Which expression represents the monthly cost of cable service?

A $15.95

B $15.95 + $3.95n$

C $3.95 + $15.95n$

D ($15.95 + 3.95)n$

GO ON

24 The altitude of an airplane on a steady descent can be found using this formula.

$$a = 27{,}000 - 1{,}500t$$

What is the altitude, a, of the airplane in feet after a time (t) of 12 minutes?

F 25,500 ft   **H** 12,000 ft

G 18,000 ft   **J** 9,000 ft

25 A manager of the sneaker department kept a record of the sizes of the most popular ladies sneakers sold in his store one day.

$5\frac{1}{2}, 5\frac{1}{2}, 6, 6, 6, 6\frac{1}{2}, 7,$

$7, 7, 7, 7\frac{1}{2}, 7\frac{1}{2}, 8, 8, 8, 8,$

$8\frac{1}{2}, 8\frac{1}{2}, 9, 9, 9\frac{1}{2}, 10$

Which measure of central tendency should he use to determine how many of the different sizes he should order for the next shipment?

A Mean   **C** Mode

B Median   **D** Range

26 Tom borrowed $124.75 from his parents to buy a new stereo system. He repaid his parents $15 a week for 5 weeks. How much does he still owe to his parents?

F $75.00   **H** $49.75

G $74.75   **J** $24.75

27 How is the graph of $y = mx + b$ affected when m remains the same and b is replaced by a different number?

A The graph is parallel to the original graph.

B The graph is perpendicular to the original graph.

C The graphs have the same y-intercepts.

D The graphs pass through the origin.

28 Triangle ABC is similar to triangle XYZ.

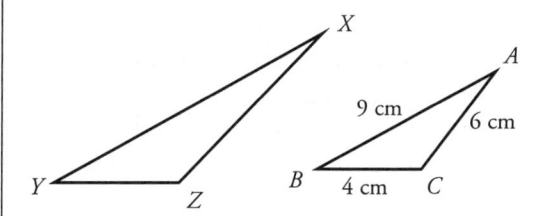

Which of the following could be the lengths of the sides of triangle XYZ?

F 8 cm, 12 cm, 16 cm

G 12 cm, 18 cm, 24 cm

H 16 cm, 24 cm, 36 cm

J 18 cm, 24 cm, 36 cm

GO ON ▶

29 The graphs of $y = x - 3$ and $x + y = 1$ are shown on the coordinate grid.

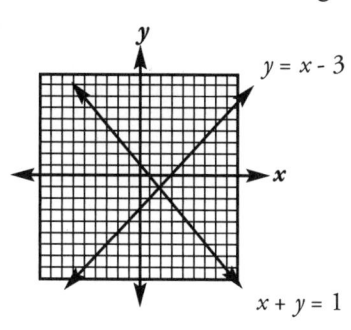

Which statement is true?

A The slopes of the lines are the same.

B The slopes of the lines have a product of 1.

C The slopes of the lines are reciprocals.

D The slopes of the lines are negative reciprocals.

30 What is the total number of faces, edges, and vertices in a triangular prism?

F 14   **H** 20

G 18   **J** 26

31 The graph below shows a relation.

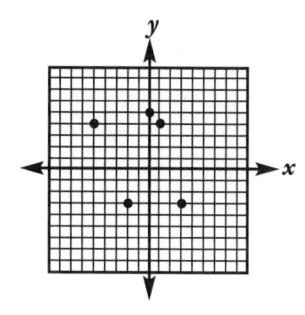

What is the range of the relation?

A {-5, -2, 0, 1, 3}

B {-3, 4, 5}

C {-5, -3, -2, 0, 1, 3, 4, 5}

D {-5, -4, -3, -2, -1, 0, 1, 2, 3, 4, 5}

32 Which expression should be next in this pattern?

$$2a - b, \; 3a + 2b, \; 4a - 4b, \; \dots$$

F $5a - 8b$ **H** $5a - 6b$

G $5a + 6b$ **J** $5a + 8b$

GO ON

33 Maria has q quarters and d dimes in her wallet. Which algebraic expression could be used to express the value of the coins in Maria's wallet?

A $0.25q + 0.10d$ **C** $q + d$

B $0.10q + 0.25d$ **D** Not Here

34 Beth paid $18.06 for a 14-pound turkey. How much should she expect to pay for a 19-pound turkey?

F $23.06   **H** $24.51

G $23.22   **J** $25.80

35 The equation of a line is $y = \frac{1}{2}x - 3$. Which graph represents this line?

A

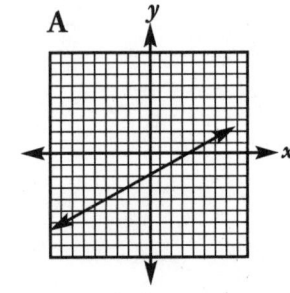

C

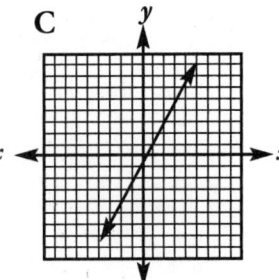

B

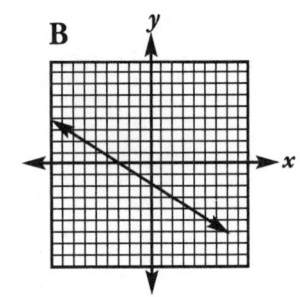

D

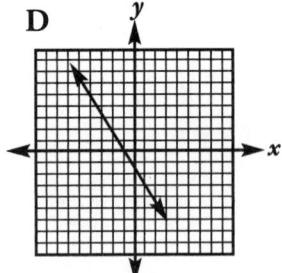

36 Mr. Brown has already saved $350 and wants to have at least $2,000 within the next year. Which expression could be used to find s, the least average amount he must save each month for the next 12 months?

F $s \geq (2000 \div 12) - 350$

G $s \leq (2000 - 350) \div 12$

H $s \geq (2000 - 350) \div 12$

J $s \leq (2000 \div 12) - 350$

GO ON

37 The ordered pair (-3, -5) is a solution of a system of equations. One of the equations in the system is $2x - 3y = 9$. Which of the following could be the other equation in the system?

A $x + y = -2$ **D** $3x + y = -4$

B $3x - y = -4$ **E** Not Here

C $x - y = -2$

38 Which statement is true about the graphs of $y = x^2 - 4$ and $y = x^2 + 1$?

F The graph of $y = x^2 - 4$ is narrower than the graph of $y = x^2 + 1$.

G The graphs of $y = x^2 - 4$ and $y = x^2 + 1$ have the same shape but different vertices located on the x-axis.

H The graphs of $y = x^2 - 4$ and $y = x^2 + 1$ have the same shape but different vertices located on the y-axis.

J The graph of $y = x^2 - 4$ opens downward and the graph of $y = x^2 + 1$ opens upward.

39 What would be the next step in solving the inequality?

$$6x + 5 - 9x \geq 14$$
$$-3x + 5 \geq 14$$
$$-3x \geq 9$$

A Divide both sides by 3. The solution is $x \geq 3$.

B Divide both sides by 3. The solution is $x \leq 3$.

C Divide both sides by -3. The solution is $x \geq -3$.

D Divide both sides by -3. The solution is $x \leq -3$.

40 Which of the following quadratic equations has roots of -5 and 3?

F $x^2 - 2x - 15 = 0$

G $x^2 - 8x - 15 = 0$

H $x^2 + 2x - 15 = 0$

J $x^2 + 8x - 15 = 0$

GO ON

41 The table shows some exponential expressions and their values.

Exponential Expression	2^4	2^3	2^2	2^1	2^0	2^{-1}
Value	16	8	4	2		

What is the value of 2^{-1}?

A -2   **C** 0

B -1   **D** $\frac{1}{2}$

42 The coordinates of the endpoints of the diameter of a circle are (4, 2) and (-2, -6). What are the coordinates of the center of the circle?

F (1, -2)   **H** (2, -4)

G (0, 0)   **J** (3, -4)

43 The diagram shows the grassy area of a track field.

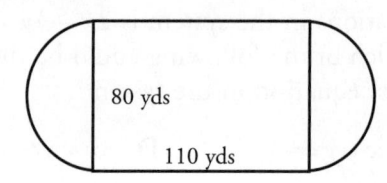

How many square yards of grass are in the track field?

A $8,800 + 1,600\pi$ yd^2

B $6,400 + 1,600\pi$ yd^2

C $8,800 + 6,400\pi$ yd^2

D $6,400 + 6,400\pi$ yd^2

44 A bag of candy contains 6 orange-flavored candies and 3 lemon-flavored candies. Glenn randomly picked one candy from the bag, then randomly picked a second piece. What is the probability that both candies picked were orange-flavored?

F $\frac{2}{3}$ **H** $\frac{5}{12}$

G $\frac{4}{9}$ **J** $\frac{10}{27}$

GO ON

45 The table shows the results of a survey of 200 high school students who were asked how they get to school in the morning.

Mode of Transportation for High School Students	
Type of Transportation	**% of Students**
Walk	20
Take the bus	40
Drive own car	15
Driven by a friend	10
Driven by parents	6
Bicycle	5
Other	4

How many degrees would be in the central angle of a circle graph for the number of students that take the bus to school?

A   40°   **C**   80°

B   72°   **D**   144°

STOP

Answer Sheet

STUDENT'S NAME																	
LAST											FIRST						MI
○	○	○	○	○	○	○	○	○	○	○	○	○	○	○	○	○	○
A	A	A	A	A	A	A	A	A	A	A	A	A	A	A	A	A	A
B	B	B	B	B	B	B	B	B	B	B	B	B	B	B	B	B	B
C	C	C	C	C	C	C	C	C	C	C	C	C	C	C	C	C	C
D	D	D	D	D	D	D	D	D	D	D	D	D	D	D	D	D	D
E	E	E	E	E	E	E	E	E	E	E	E	E	E	E	E	E	E
F	F	F	F	F	F	F	F	F	F	F	F	F	F	F	F	F	F
G	G	G	G	G	G	G	G	G	G	G	G	G	G	G	G	G	G
H	H	H	H	H	H	H	H	H	H	H	H	H	H	H	H	H	H
I	I	I	I	I	I	I	I	I	I	I	I	I	I	I	I	I	I
J	J	J	J	J	J	J	J	J	J	J	J	J	J	J	J	J	J
K	K	K	K	K	K	K	K	K	K	K	K	K	K	K	K	K	K
L	L	L	L	L	L	L	L	L	L	L	L	L	L	L	L	L	L
M	M	M	M	M	M	M	M	M	M	M	M	M	M	M	M	M	M
N	N	N	N	N	N	N	N	N	N	N	N	N	N	N	N	N	N
O	O	O	O	O	O	O	O	O	O	O	O	O	O	O	O	O	O
P	P	P	P	P	P	P	P	P	P	P	P	P	P	P	P	P	P
Q	Q	Q	Q	Q	Q	Q	Q	Q	Q	Q	Q	Q	Q	Q	Q	Q	Q
R	R	R	R	R	R	R	R	R	R	R	R	R	R	R	R	R	R
S	S	S	S	S	S	S	S	S	S	S	S	S	S	S	S	S	S
T	T	T	T	T	T	T	T	T	T	T	T	T	T	T	T	T	T
U	U	U	U	U	U	U	U	U	U	U	U	U	U	U	U	U	U
V	V	V	V	V	V	V	V	V	V	V	V	V	V	V	V	V	V
W	W	W	W	W	W	W	W	W	W	W	W	W	W	W	W	W	W
X	X	X	X	X	X	X	X	X	X	X	X	X	X	X	X	X	X
Y	Y	Y	Y	Y	Y	Y	Y	Y	Y	Y	Y	Y	Y	Y	Y	Y	Y
Z	Z	Z	Z	Z	Z	Z	Z	Z	Z	Z	Z	Z	Z	Z	Z	Z	Z

SCHOOL:

TEACHER:

FEMALE ○ **MALE** ○ **GRADE:**

BIRTH DATE

MONTH	DAY		YEAR	
Jan ○	0	0	0	0
Feb ○	1	1	1	1
Mar ○	2	2	2	2
Apr ○	3	3	3	3
May ○		4	4	4
Jun ○		5	5	5
Jul ○		6	6	6
Aug ○		7	7	7
Sep ○		8	8	8
Oct ○		9	9	9
Nov ○				
Dec ○				

Answer Sheet

Fill in the circle for each multiple-choice answer. Write the answers to the open-ended questions in the space provided on the test.

Social Studies Practice Test

SA Ⓐ Ⓑ Ⓒ Ⓓ	9 Ⓐ Ⓑ Ⓒ Ⓓ	18 Ⓕ Ⓖ Ⓗ Ⓙ	27 Ⓐ Ⓑ Ⓒ Ⓓ	36 Ⓕ Ⓖ Ⓗ Ⓙ
1 Ⓐ Ⓑ Ⓒ Ⓓ	10 Ⓕ Ⓖ Ⓗ Ⓙ	19 Ⓐ Ⓑ Ⓒ Ⓓ	28 Ⓕ Ⓖ Ⓗ Ⓙ	37 Ⓐ Ⓑ Ⓒ Ⓓ
2 Ⓕ Ⓖ Ⓗ Ⓙ	11 Ⓐ Ⓑ Ⓒ Ⓓ	20 Ⓕ Ⓖ Ⓗ Ⓙ	29 Ⓐ Ⓑ Ⓒ Ⓓ	38 Ⓕ Ⓖ Ⓗ Ⓙ
3 Ⓐ Ⓑ Ⓒ Ⓓ	12 Ⓕ Ⓖ Ⓗ Ⓙ	21 Ⓐ Ⓑ Ⓒ Ⓓ	30 Ⓕ Ⓖ Ⓗ Ⓙ	39 Ⓐ Ⓑ Ⓒ Ⓓ
4 Ⓕ Ⓖ Ⓗ Ⓙ	13 Ⓐ Ⓑ Ⓒ Ⓓ	22 Ⓕ Ⓖ Ⓗ Ⓙ	31 Ⓐ Ⓑ Ⓒ Ⓓ	40 Ⓕ Ⓖ Ⓗ Ⓙ
5 Ⓐ Ⓑ Ⓒ Ⓓ	14 Ⓕ Ⓖ Ⓗ Ⓙ	23 Ⓐ Ⓑ Ⓒ Ⓓ	32 Ⓕ Ⓖ Ⓗ Ⓙ	41 Ⓐ Ⓑ Ⓒ Ⓓ
6 Ⓕ Ⓖ Ⓗ Ⓙ	15 Ⓐ Ⓑ Ⓒ Ⓓ	24 Ⓕ Ⓖ Ⓗ Ⓙ	33 Ⓐ Ⓑ Ⓒ Ⓓ	42 Ⓕ Ⓖ Ⓗ Ⓙ
7 Ⓐ Ⓑ Ⓒ Ⓓ	16 Ⓕ Ⓖ Ⓗ Ⓙ	25 Ⓐ Ⓑ Ⓒ Ⓓ	34 Ⓕ Ⓖ Ⓗ Ⓙ	
8 Ⓕ Ⓖ Ⓗ Ⓙ	17 Ⓐ Ⓑ Ⓒ Ⓓ	26 Ⓕ Ⓖ Ⓗ Ⓙ	35 Ⓐ Ⓑ Ⓒ Ⓓ	

English Language Arts Practice Test

1 Ⓐ Ⓑ Ⓒ Ⓓ	7 Ⓐ Ⓑ Ⓒ Ⓓ	13 Ⓐ Ⓑ Ⓒ Ⓓ	19 Ⓐ Ⓑ Ⓒ Ⓓ	25 Ⓐ Ⓑ Ⓒ Ⓓ
2 Ⓕ Ⓖ Ⓗ Ⓙ	8 Ⓕ Ⓖ Ⓗ Ⓙ	14 Ⓕ Ⓖ Ⓗ Ⓙ	20 Ⓕ Ⓖ Ⓗ Ⓙ	26 Ⓕ Ⓖ Ⓗ Ⓙ
3 Ⓐ Ⓑ Ⓒ Ⓓ	9 Ⓐ Ⓑ Ⓒ Ⓓ	15 Ⓐ Ⓑ Ⓒ Ⓓ	21 Ⓐ Ⓑ Ⓒ Ⓓ	27 OPEN-ENDED
4 Ⓕ Ⓖ Ⓗ Ⓙ	10 Ⓕ Ⓖ Ⓗ Ⓙ	16 Ⓕ Ⓖ Ⓗ Ⓙ	22 Ⓕ Ⓖ Ⓗ Ⓙ	
5 Ⓐ Ⓑ Ⓒ Ⓓ	11 Ⓐ Ⓑ Ⓒ Ⓓ	17 Ⓐ Ⓑ Ⓒ Ⓓ	23 Ⓐ Ⓑ Ⓒ Ⓓ	
6 Ⓕ Ⓖ Ⓗ Ⓙ	12 Ⓕ Ⓖ Ⓗ Ⓙ	18 Ⓕ Ⓖ Ⓗ Ⓙ	24 Ⓕ Ⓖ Ⓗ Ⓙ	

Mathematics Practice Test

1 Ⓐ Ⓑ Ⓒ Ⓓ	11 Ⓐ Ⓑ Ⓒ Ⓓ	21 Ⓐ Ⓑ Ⓒ Ⓓ	31 Ⓐ Ⓑ Ⓒ Ⓓ	41 Ⓐ Ⓑ Ⓒ Ⓓ
2 Ⓕ Ⓖ Ⓗ Ⓙ	12 Ⓕ Ⓖ Ⓗ Ⓙ	22 Ⓕ Ⓖ Ⓗ Ⓙ	32 Ⓕ Ⓖ Ⓗ Ⓙ	42 Ⓕ Ⓖ Ⓗ Ⓙ
3 Ⓐ Ⓑ Ⓒ Ⓓ	13 Ⓐ Ⓑ Ⓒ Ⓓ	23 Ⓐ Ⓑ Ⓒ Ⓓ	33 Ⓐ Ⓑ Ⓒ Ⓓ	43 Ⓐ Ⓑ Ⓒ Ⓓ
4 Ⓕ Ⓖ Ⓗ Ⓙ	14 Ⓕ Ⓖ Ⓗ Ⓙ	24 Ⓕ Ⓖ Ⓗ Ⓙ	34 Ⓕ Ⓖ Ⓗ Ⓙ	44 Ⓕ Ⓖ Ⓗ Ⓙ
5 Ⓐ Ⓑ Ⓒ Ⓓ	15 Ⓐ Ⓑ Ⓒ Ⓓ	25 Ⓐ Ⓑ Ⓒ Ⓓ	35 Ⓐ Ⓑ Ⓒ Ⓓ	45 Ⓐ Ⓑ Ⓒ Ⓓ
6 Ⓕ Ⓖ Ⓗ Ⓙ Ⓚ	16 Ⓕ Ⓖ Ⓗ Ⓙ	26 Ⓕ Ⓖ Ⓗ Ⓙ	36 Ⓕ Ⓖ Ⓗ Ⓙ	
7 Ⓐ Ⓑ Ⓒ Ⓓ Ⓔ	17 Ⓐ Ⓑ Ⓒ Ⓓ	27 Ⓐ Ⓑ Ⓒ Ⓓ	37 Ⓐ Ⓑ Ⓒ Ⓓ Ⓔ	
8 Ⓕ Ⓖ Ⓗ Ⓙ	18 Ⓕ Ⓖ Ⓗ Ⓙ	28 Ⓕ Ⓖ Ⓗ Ⓙ	38 Ⓕ Ⓖ Ⓗ Ⓙ	
9 Ⓐ Ⓑ Ⓒ Ⓓ	19 Ⓐ Ⓑ Ⓒ Ⓓ	29 Ⓐ Ⓑ Ⓒ Ⓓ	39 Ⓐ Ⓑ Ⓒ Ⓓ	
10 Ⓕ Ⓖ Ⓗ Ⓙ	20 Ⓕ Ⓖ Ⓗ Ⓙ	30 Ⓕ Ⓖ Ⓗ Ⓙ	40 Ⓕ Ⓖ Ⓗ Ⓙ	